A Best of
Creative Trai‗‗‗‗‗‗‗‗‗‗‗‗‗‗‗‗‗‗‗‗‗‗‗‗‗‗‗*ook*

P9-EMN-513

POWERFUL AUDIOVISUAL TECHNIQUES

101
Ideas to
Increase the Impact
and Effectiveness
of Your Training

■ ■ ■

By Bob Pike with Julie Tilka

Lakewood Publications
A Maclean Hunter Company

Quantity Sales

Most Lakewood books are available at special quantity discounts when purchased in bulk by companies, organizations and special-interest groups. Custom imprinting or excerpting can also be done to fit special needs. For details contact Lakewood Books.

■ ■ ■

LAKEWOOD BOOKS
50 South Ninth Street
Minneapolis, MN 55402
(800) 707-7769 or (612) 333-0471
FAX (612) 340-4819

Publisher: Philip G. Jones
Editors: Bob Pike with Julie Tilka
Production Editor: Julie Tilka
Production: Carol Swanson and Pat Grawert
Cover Designer: Barb Betz, Betz Design

10 9 8 7 6 5 4 3 2 1

Lakewood Publications, Inc. publishes *TRAINING Magazine; Training Directors' Forum Newsletter; Creative Training Techniques Newsletter; Technology For Learning Newsletter; Potentials In Marketing* Magazine, *Presentations* Magazine; and other business periodicals, books, research and conferences.

Bob Pike, Creative Training Techniques International, 7620 W. 78th St., Edina, MN 55439, (612) 829-1960, FAX (612) 829-0260.

ISBN 0-943210-66-6

Contents

Foreward

This book, *Powerful Audiovisual Techniques*, is one in a series drawn from the best content of *Creative Training Techniques Newsletter*. The newsletter was conceived in 1988 by editor and internationally known trainer Bob Pike to be a one-stop resource of practical "how-tos" for trainers. The idea was (and still is) to provide timely tips, techniques, and strategies that help trainers with the special tasks they perform daily.

When the newsletter began, it was largely fueled by Bob's 20 years of experience in the field and by the best ideas shared by the trainers (more than 50,000 in all) who had attended his Creative Training Techniques seminars. As the newsletter grew in popularity, it also began to draw on ideas submitted by its readers. Today, the newsletter continues to search out creative approaches from the more than 200 seminars Bob and the other Creative Training Techniques trainers conduct every year and from the newsletter readers.

But no matter where the insights come from, the goal of the newsletter remains the same: To provide trainers a cafeteria of ideas they can quickly absorb, and then choose the ones that best suit their special needs.

This series of books represents the best ideas from *Creative Training Techniques Newsletter's* six years of publication. It is our hope we've created a valuable resource you'll come back to again and again to help address the unique challenges you face in your job daily.

Sincerely,
The Editors

Introduction

We live in the age of entertainment. Forty years ago we had three TV stations, all broadcasting in black and white. Today we grab the remote and "channel surf" through 20 times more channels until something — if anything — grabs our attention. Just like today's media-savvy, sophisticated TV viewers, a training audience learns best when we capture and keep its attention.

Of course, flashy audiovisuals can never replace solid content, but more than ever, being effective at using audiovisuals is vital to reaching and engaging our audiences. *Powerful Audiovisual Techniques* gives you tips, tactics, and techniques for using a wide variety of visuals more effectively.

We encourage you to "mix it up." Using several formats keeps your presentations fresh and interesting for your audience. Teaching the same content over and over can be boring, but changing the visual approach used from one class to another can help keep you — the presenter — fresh, too.

And as you develop a training program, keep in mind the importance of using several different formats. That way, if a flip chart isn't available, you can use transparencies, or if your last overhead bulb burns out, you can use slides or flip charts. You don't want to be handicapped by being dependent on only one visual approach.

Here's a brief description of the tips you'll find in each chapter:

• *Chapter One* shows you how to make the most of

your transparencies.

- *Chapter Two* gives you tips for using flip charts.
- *Chapter Three* offers techniques to make your slide presentations more engaging.
- *Chapter Four* shows you how to use videotapes to their maximum.
- *Chapter Five* includes ideas for using a variety of audiovisuals more effectively, and how to set the stage for your presentation through lighting, sound, and room set-up guidelines.
- *Chapter Six* provides activities that use audiovisuals as a learning tool.

Powerful Audiovisual Techniques can be used in a variety of ways. Browse through it, as you might a cookbook, and look for a method that strikes your fancy. Go to the section that contains the visual approach you use most often and look for a fresh approach. Or go to a section and look for some interesting ways to use visuals that are not a common part of your repertoire.

You might just choose to read *Powerful Audiovisual Techniques* from cover to cover. One suggestion: Don't try to implement every idea at once. Select several, use them, then as they become comfortable to you, come back and look for more. It's better to implement two or three new ideas well then to try to implement 10 with far less effectiveness.

Bob Pike

Chapter One Tips: Tranparencies

C olored acetates are a versatile, inexpensive way to highlight transparencies, says E. Douglas Pratt, associate trainer with the Child Welfare Institute in Atlanta. Pratt cuts the acetate into shapes of various sizes, then lays a rectangle over a section of the transparency he wants to emphasize, and uses a triangle as a pointer to emphasize a word or line. Because the shapes are loose, they can be easily moved to other sections of an overhead. They also pack and travel easily, Pratt says.

Colored Acetates

Versatile, Inexpensive Way to Highlight Transparencies

- Cut the acetate into shapes of various sizes.
- Lay a shape over the section of the transparency you want to emphasize.
- Use a triangle as a pointer to a word or line.
- The shapes can easily be moved to another section of an overhead. They pack and travel easily, too.

2

Acetate report covers add color to overheads

Though colored transparencies add zip and emphasis to an overhead presentation, Billie Lott, trainer for Health and Welfare in Caldwell, ID, finds colored-acetate report covers are a cheaper, acceptable alternative.

Lott cuts each cover in half to get two 9 x 12-inch sheets. Then she either places the report cover on top of the clear transparency, or cuts the report cover to fit over certain portions for a highlighting effect. Yellow, blue, green, and neon colors work best, and light pink and blue can be overlaid to create lavender.

Red report covers are not translucent, so Lott cuts them into the universal symbol for "do not" (circle with a diagonal slash). She uses the symbols to emphasize changes or incorrect information. For example, she places the red symbol over a transparency that shows the old instructions that have been deleted from a manual.

Cardstock and cookie cutters keep attendees' eyes and minds focused on the most important parts of overhead projections.

When Donald Blum, manager of education and training at Ford Motor Co., creates a transparency he also creates an overlay by tracing cookie cutters of various shapes on cardstock and cutting them out. The holes are positioned so that when the cardstock is placed over the transparency key points are visible through the holes.

Once a transparency has been discussed in detail, Blum places the overlay on it, leaving only the most important items visible. He then focuses the discussion on those points by saying, for example, "Let me direct your attention to the key point in the star."

Cookie cutters are available in hundreds of shapes, he says, so finding shapes that relate to a particular lesson is often possible.

3

Cookie cutter cutouts highlight important points

4

Recycle a slide show: Turn it into an overhead program

Transferring overheads to slides, doesn't need to be expensive or require specialized equipment to achieve professional results.

To convert an overhead presentation into a slide presentation, David Van Doren, manager of development for Boehringer Ingelheim, Southbury, CT, suggests simply using a 35mm camera with slide film. For example, for a presentation illustrating workers and activities at a hospital, he photographs a variety of scenes at a local hospital, and shoots slides of appropriate overheads.

Van Dorn says a polished audiovisual presentation can be produced for about $25.

U sing small Post-it Notes makes it easier to separate and pick up transparencies, says John Schamel, a trainer with the National Education Association of New York, Elmira, NY. Schamel arranges the Post-its in a step fashion on either edge of his transparencies, making it easy to pick up each sheet in order. He also numbers the notes, so there is no question about sequence.

Because Post-its peel off the transparencies easily, it's easy to change their order from presentation to presentation. And since Post-its come in a variety of colors, transparencies can be quickly color-coded as well, Schamel says.

Post-it Notes make picking up overheads easy

Arranging Post-it Notes in a step fashion on either edge of the transparencies helps make them:

- Easier to pick up.
- Easier to keep in order.
- Easier to color code.

6

Make your pointer tip easier to see

Here's a tip on tips from Mark Sullivan, senior engineer of PP&L of Berwick, PA. If you use a pointer with an overhead projector, and you point at the screen rather than at the transparency itself, a pointer with a shiny metal tip can cause a distracting glare. Avoid that by wrapping masking tape over the tip and coloring it with a yellow marker. That focuses attention on the pointer tip without the bothersome glare.

Here's a few simple tricks for keeping your overhead presentation on track.

■ Carol Lauffer, field director of the Mitten Bay Girl Scout Council, says Post-it Notes are a great way to cover parts of an overhead you want to slowly reveal and also act as ideal "cue cards." Lauffer jots notes on each Post-it to remind herself what to say before revealing each point.

■ For bolstering a trainer's self-confidence when presenting new material or making lengthy presentations, Katherine Vyse of Scotiabank, West Toronto, Ontario, suggests this tip:

In the lower right corner of each overhead tranparency frame, write a key word or phrase that reminds you of the content of your next transparency. This makes it easy to unobtrusively glance at the frame to ensure that you are on the right track. It's especially useful, too, if your group sidetracks you momentarily.

7

Tricks to keep your presentation on track

8

Purposely flawed overheads keep trainees on their toes

Overhead transparencies that deliberately contain wrong information keep trainees on their toes, says Jane Grosslight, program administrator at Florida State University.

She says the inaccuracies encourage participants to pool their knowledge and use whatever resources available to identify and correct the inaccurate information on the overhead.

As a variation, Grosslight splits the class into small groups that generate corrections and then compare suggestions.

Leaving on the light of an overhead projector while changing transparencies is distracting. Yet the incessant click, click, click of turning the overhead on and off repeatedly is also distracting. The problem is eliminating the distraction, not turning off the light. Here are a couple of solutions:

■ Trim a piece of cardboard or a file folder so that it's about 11 inches x 11 inches. (The normal stage on an overhead is 10 inches x 10 inches.) Tape two or three pennies to the top of the cardboard. Now slide the cardboard under your transparency. Move the cardboard down as you reveal your information; the pennies will keep it in place without you having to hold it. When you're finished, slide the cardboard up so it completely covers the lighted stage. Remove your transparency and place the next one on top of the cardboard.

■ Laura Wassmer, training administrator for Hallmark Cards, Kansas City, MO, keeps her overhead projector from showing "white" on the screen between transparencies by taping a piece of cardboard to the top of the projector head. She flips it up when she's showing the overhead or down to darken the screen.

9

Eliminate 'white screen' between overheads

10

Projecting onto a whiteboard makes forms reusable

Try using your overhead to project the image of a standard form when teaching people to fill them out.

■ Kathleen Slater, a division supervisor for Progressive Insurance, projects a form onto a whiteboard so participants can practice filling it out properly and then erase their responses, clearing the form for the next person. This technique allows for continuous practice and enables a trainer to change the form at will.

■ Project transparencies onto a whiteboard where you can write extra information and add emphasis, suggests Pat Stewart of Loblaws Supermarkets, Toronto. Stewart says projecting on a whiteboard is especially useful for outlining corporate heirarchies and for demonstrating the steps in filling out forms.

Consider investing in a remote control device for your overhead projector to improve your mobility around a class. David Pomeroy, who does technical training with gas fireplaces, bought a remote control for his projector so he can turn the overhead on or off from anywhere in the room when he wants to make a point, while strolling through the audience.

Remote control allows greater mobility

12

Three-ring binder keeps overheads organized

The 200 transparencies that director of training Roseanne Brennan, Howard Johnson Franchise Systems, uses during a three-day course were difficult to organize and store. To make them more manageable, Brennan placed each transparency in a clear-plastic sheet protector, then snapped them all into a three-ring binder.

When presenting, she keeps the binder open, takes out each sheet protector with the transparency still inside, and places it on the overhead projector. She doesn't recommend nonglare protectors; they cloud the image.

In addition, hand-colored cartoon transparencies don't get smeared, and the covers of those used for exercises can be written on with washable markers and wiped clean.

She numbers the transparencies to keep them organized. Her seminar includes seven sections, so she marks her transparencies I-1, I-2, etc., with sections separated by dividers. Each transparency is also marked to indicate the page number in the participant manual where the information can be found.

The most common method for selectively revealing information on overhead transparencies is to cover the transparency with a sheet of paper and to slide the paper down to show each new point. An alternative method is to cut a file folder into strips, leaving one edge uncut. Tape the uncut edge along one edge of the transparency, and as you reveal each new piece of information, lift the appropriate cut section of the file folder.

One advantage of this method is that information can be revealed randomly. You may need to jot some notes on the folder strips, since you will be unable to read the information through the folder.

13

Revelation method lets you selectively show points

14

Highlighting technique adds color to overheads — cheap

Highlighting is a tactic commonly used with 35mm slides, where a main point is displayed in bright color while other points remain in a background color.

You can use the same technique with your transparencies. Two types of transparency film are required: Negative-image film (such as Thermaview made by Arkwright) and write-on film (such as Colorburst made by Labelon).

The negative-image film allows you to make a transparency so the points are visible when projected, but the rest of the screen remains black because the transparency film blocks the light. Place your negative-image transparency on the overhead bed, but before turning the bulb on, place a sheet of highlighting film (which is royal blue) on top of the first transparency.

When the overhead is turned on, your points will be dimly visible because of the low contrast between black and blue. As you discuss each point, take the highlight pen and run it over the blue film. The pen turns the film a brilliant yellow and makes that point clearly visible.

For information on the transparency films contact Labelon (800) 428-5566 and Arkwright (800) 638-8032.

Your transparencies will have greater impact and a more professional appearance if you can remember the following guidelines:

• Screen width = 1/6 of the room length (a room 36 feet long would require a 6-foot-wide screen)

• Front row seats no closer than two screen widths from screen. (A 6-foot-wide screen would have the first row 12 feet away.) This allows for positioning the overhead projector without putting it right on top of participants.

• Last row of seats no further than six screen widths. (A 6-foot-wide screen means the last row should be no further than 36 feet away.)

• No row of seats wider than its distance from the screen. (12 feet from the screen, 12 feet wide; 24 feet from the screen, 24 feet wide.)

• Bottom of the screen at least 42 inches from the floor. (Makes the bottom of the screen easier to read when participants are seated behind other people.)

• Projected image completely fills the screen. (Move your projector to make sure this occurs.)

15

Overheads have greater impact with room set-up guidelines

16

Content tips ensure trainees get the point

These transparency content guidelines will ensure participants get the most out of each transparency:

- Focus on one idea per visual.
- Highlight key points.
- Use check marks, bullets, or boxes to highlight nonsequential points.
- Make sure your visuals are "visual" whenever possible — use charts, graphs, clip art.
- Avoid copying pages out of a book and turning them into visuals.
- Include on the visual only information you intend to discuss.
- Visuals support the spoken word; visual are not for reading word-for-word.
- Limit work area on original to 7 1/2 x 9 1/2 inches.
- Minimum letter size = 1/4 inch.
- Vary size of lettering to demonstrate relative importance of information.
- Use same type style on a series of related visuals.
- Space letters optically, not mechanical.
- Between each line of type, leave a space equal to the height of your uppercase lettering.
- Maximum six words per line, six lines per visual.

Tired of fumbling in the dark while trying to align your transparencies? When using an overhead projector, Laura Wassmer, Ambassador training administrator for Hallmark Cards, tapes a pencil at the top edge of the light platform so she can quickly drop on a transparency and align it perfectly every time — even in the dark.

Tape pencil to top edge of light platform

18

Use an ounce of prevention: Plan your presentation

Using overhead transparencies can be a deceivingly simple means of presenting information to trainees. Overheads are versatile, simple, and inexpensive to produce, and add impact to any session, but there are some elementary rules of preparation to consider. Here's a refresher course on some of those basic rules:

• Number your transparencies so that if you drop them or otherwise shuffle them, you can sort them out easily.

• Analyze your transparencies and decide which could be eliminated if necessary. Put a red star in the corner of the expendable ones and a matching star in the appropriate section of your lecture notes. There may be times when you need to shorten your presentation and these specially marked transparencies will give you the flexibility to do so gracefully.

• Arrive early to oversee the set up procedures. Verify for yourself that everything is ready; don't rely on someone else's word.

• Carry an extension cord — just in case. Also, carry a kit of other supplies such as tape, scissors, and tacks for emergency situations.

• Set up and test the equipment. Place the first transparency on the light table and check for proper screen placement and image distortion. Lay out your transparencies in order.

• Test the room's lighting with a transparency on the light table. Keep the lights as bright as possible while maintaining image clarity. The more light you have, the better you can control the audience.

• Have a contingency plan. What if none of your equipment works? You should be able to give a good presentation just referring to your transparencies as lecture notes. Don't leave malfunctioning equipment in place. Put it out of sight and go on with the presentation without apologies.

• Practice, practice, practice. Even though you've carefully marked your transparencies and their corresponding places in the lecture notes, it's easy to get confused when the presentation is in progress.

19

Glossy magazines inexpensive source for color overheads

David Smith, a trainer with Seeman Furniture, finds he can produce great transparencies in full color from glossy magazines. Many magazines — *Time, Newsweek, Life* — are printed on what is called *coated stock.* With these magazines you can transfer any color picture to a transparency using these four steps:

1. Thinly coat with rubber cement the magazine picture you want to transfer.

2. Glue the picture to a piece of clear transparency film.

3. After the rubber cement dries, soak the transparency in warm, soapy water.

4. After the magazine paper is thoroughly soaked, peel it off, leaving only the ink attached to the plastic.

Chapter Two Tips: Flip Charts

20

Preparing text on a computer lends a professional look

Flip charts can take on a professional look if you prepare the text for your pages on a computer first, print the images, then project and trace them onto flip-chart paper, says Janet Braud, marketing training and communications manager for Herff Jones Inc., Montgomery, AL.

Braud uses a personal computer to first create text in a page layout software, such as Quark Express or Aldus PageMaker, so she can easily see how everything will fit on individual pages and make any necessary corrections. Simple graphics that can be quickly traced can also be inserted at this point. Sans serif font types such as Helvetica or Tektron, she says, are ideal to use because the straight edges are extremely easy to read and trace.

Once she has everything in place, Braud prints the documents and projects them onto the flip chart with an opaque projector. Last-minute spacing adjustments can be made by simply moving the flip-chart easel closer to or farther away from the projector. Braud then traces the images and text with markers or, if time is short, with a pencil so she can later write over the image in marker.

A "For Your Information" list can encourage participants to share useful resources and tips with each other. Post a piece of flip-chart paper titled "FYI" at the beginning of your workshop and encourage participants to write on it throughout the course. They might contribute a book title, resource, or a job practice that's worked well for them. After the program, reprint the list and send it to participants with other follow-up material.

Make your own contributions to the "FYI" list by putting related books, news clippings, and articles of interest on a resource table near the list.

21

FYI list encourages trainees to share resources

22

Chalkboard compass creates flawless circles

To make quick and flawless circles on flip charts or whiteboards, Mike Davis, facilitator with Zenger-Miller FrontLine Leadership training programs, uses a chalkboard compass purchased from a school supply store. He simply uses a rubberband to attach a marker to the arm where the chalk would usually go, and can then make circles up to 48 inches in diameter. Davis says the compasses typically sell for $10 to $20.

W henever an issue comes up that needs to be given more attention — either because that material will be addressed later or because the trainer needs to find out something for the group — Kim Dean, marketing training and communications manager for Herff Jones Inc., Montgomery, AL, puts it in the "parking lot," a flip-chart page designed as shown below. Before the class is over, she tells attendees, the parking lot will be empty.

This reminds the trainer and the group, Dean says, that those issues need to be resolved, and makes participants feel their questions are important.

24

**Hiding
flip-chart
info
piques
trainee
curiosity**

In courses with a lot of prepared flip-chart pages, Beverly Peavey, training and development manager for Sprint, Lenexa, KS, tapes the pages to the walls, but pins the bottoms of the pages to the tops so the material is hidden. When she is ready to reveal the information, Peavey simply unpins the page. She says the strategy saves her time and energy she might otherwise spend wrestling with a flip-chart easel, allows her to walk among the group as she moves from page to page, and also serves as a constant review tool since the material is hung for all to see throughout the course.

C onsider using several flip-chart easels (up to five or six) across the front of the classroom if and when any of the following conditions apply:

• You must develop a continuous, uninterrupted flow of ideas, and you can't take time to post pages on the wall.

• The room is big, requiring your printing to be quite large.

• You have a large, complex chart or diagram to display.

25

Use several flip charts when special conditions exist

26

The 'Six T's' for effective flip-charting

Rosetta Nelson, an instructor development specialist for Florida Power and Light, recommends these six T's for effective flip- charting: Touch, Turn, Talk, Tape, Tear, Tape.

Touch the item you want the group to focus on.

Turn to the audience.

Talk to the audience not the flip-chart.

Tape goes on the flip-chart page.

Tear off the flip-chart page.

Tape it to the wall.

W hen no overheads or wall-boards are available during on-site technical training, Sandra Lott, senior operations training specialist at Loblaws Supermarkets in Toronto, substitutes pre-prepared, laminated flip-chart pages. Lott uses erasable markers on the sheets, so they can be used again and again, minimizing cost and preparation time.

■ Marsha Stevenson, Clipper Petroleum director of training, also makes her flip charts reusable. She places a piece of protective film over the base flip chart, which allows her to write on and highlight points that can be erased with a damp cloth and leave the chart ready to use again just as a transparency might be.

To get even more mileage out of a single lamination, put two drawings back-to-back, then laminate them.

27

Laminating flip-chart pages increases their lifespans

28

Raise your flip-chart's height for greater crowd visibility

We've been told by experts that the top two-thirds of a flip chart is the only useful part, largely because in most rooms people's heads are blocking the bottom third.

Larry Ecton, a trainer with Summit Electric Supply, who looks at things from the vantage point of being 6 feet 4 inches tall, decided to save wear and tear on his back by adding 10-inch extensions to the legs of his flip-chart easel. (His were custom-made by his company's modification shop, from inexpensive electrical conduit.)

The unanticipated benefit: The entire chart is easy to see from everywhere in most rooms. And Ecton is relieved to report he has no more backaches from bobbing as he records.

C olored transparencies or cellophane provide a quick way to add color to your prepared flip charts, says Carol J. Webb, a staff development specialist at Clemson University. Markers leave streaked color and are time-consuming to use over a large area, Webb says. Pieces of colored cellophane positioned with double-stick tape are an easier solution.

29

Cellophane sheets add color to flip charts

30

'Compliment sheets' remind students to praise each other

In addition to whatever positive feedback you are giving participants for energetic participation or insightful answers or the like, allow them to also congratulate and compliment each other.

Harold Camacho, an instructor with SVG Lithography Systems, Wilton, CT, asks each participant to write his or her name on a piece of flip-chart paper, and then hangs the pages around the room. During breaks or directly after team efforts, participants may write a complimentary note about another participant on a Post-it Note and attach it to that person's poster. Attendees take their posters home with them at the session's conclusion.

Here's more proof that you don't have to be an artist to make flip charts look professional. Taray Klein, a computer specialist for the USDA National Finance Center in New Orleans, projects a transparency she wants to use on a flip chart before her session begins. She then lightly traces the model image in pencil. During the session Klein uses a marker to draw over the traced image. She achieves a look of spontaneity, and her flip charts are well-planned and clearly illustrated as a result of her preparation.

31

Tracing drawings lets you clearly illustrate points

32

Two ways to carry unwieldy flip-chart pages

A proven way to protect and transport posters and flip charts you use repeatedly is to roll them up. But if you frequently travel by air and want to check them as baggage, you may want to build a carrying case.

■ John Kidwell, manager of training for Mazzio's Corp., built an inexpensive, sturdy transport tube for his flip charts by gluing an end-cap onto one end of a section of plastic PVC plumbing pipe and screwing a lid onto the other end. All materials are available at hardware and plumbing supply stores.

■ Edith Iwata, a trainer with Intel, uses a different approach. She uses a skirt hanger (the kind with the rubberized grips to prevent slippage), folds the pages once loosely, and then clips them to the hanger at their open end. She puts her markers and masking tape in a plastic bag and also attachs it, creating an easy method for transporting pages and materials.

Chapter Three Tips: Slides

33

Yellow filter adds color to slides

It's easy to brighten up a presentation by turning black-and-white 35mm slides of text into blue and yellow slides, says David Drehmer, an associate professor at DePaul University, Chicago. Just project old slides on a screen and photograph them using Ektachrome film and a dark yellow filter on your lens. Ask your photo developer to process the film in C-41 chemistry. On the new slides, what was white becomes vivid yellow and what was black appears blue.

A program that lasts four days or longer gives you time to develop a high-impact, personalized closing. On the first or second day designate someone to take 35mm slides of all of your participants. Get action shots — participants engaged in skits, role-playing, involved in small-group discussion, and on break. Ask for a variety of angles — full shots showing the classroom or several people working together, as well as tight shots showing one individual concentrating or chatting. Send the slides out for overnight developing, so you have time to review the slides and put together an entertaining show.

At the session's final lunch or during a coffee break, you can show the slides, perhaps with a simple musical background. Integrate some preplanned slides related to the course content, and once you've established a standard format it's easy to pop in students' slides unique to each session. That preparation minimizes the extra effort required on the first day and results in a memorable closing.

34

Slide show featuring trainees creates a high-impact closing

35

Typeface selection greatly improves slide readability

When putting together a slide or transparency, a simple change of typeface can improve the readability of your visual.

Trainers often use a serif typeface, an example of which is given below at left. The curves that embellish the tips of the letter make a serif typeface more difficult to take in at a glance. Simply changing the typeface to a sans serif (without the fancy finishes) dramatically increases visibility and readability without changing the size of the typeface. Helvetica, Futura (used on this page for the tip headline and number), Optima, and Univers are samples of sans serif (also called Gothic) typefaces.

Serif   Sans Serif

To give trainees a mental break during technical presentations that use lots of slides, Ann Cichon-Bales, a training consultant based in Vallejo, CA, intersperses scenic slides of a waterfall, scenes from Yosemite National Park, a doe in the woods, or a rushing river. The scenic slides help participants relax and enjoy the opportunity to "get out of the classroom" for a few minutes, she says.

36

Scenic slides help trainees take a mental break

37

Planning, executing tips for better slide shows

Single-projector slide shows are inexpensive to produce, can be assembled quickly, and are easy to update. But their simplicity can lull you into a sense of complacency. Here are five steps to better slide shows that can make noticeable improvements in your next program:

1. *Analyze the audience.* Who, exactly, will be viewing the production? What is their educational background? Their level of experience? Establishing "who" at the outset guarantees your program will be on target, and you may find you'll have to produce a few different versions to be really effective.

2. *Plan before you start.* The "what" part of the presentation is its message, and you must consider everything from specific terms that need to be covered to your desired reaction from the audience. The purpose of the visual design of a training presentation is to capture the viewer's attention with the first slide and not let go until the last one. Developing a verbal and visual plan before you begin active production will save time and money.

3. *Keep it short.* Resist the urge to create an epic. Even the best single-projector slide show will begin to lose viewer interest after about 15

minutes. If the subject demands that your program be longer, consider breaking it into several smaller modules.

4. *Write as you talk.* Use short, simple sentences when preparing the script for your show and keep it as lively and informal as possible. Remember, you're talking to a group of people, not writing a formal thesis for a doctorate. When you've finished the first draft, tape record a reading of it and play it back. Does it sound natural, as if you're speaking instead of reading?

5. *Create a lively soundtrack.* Music and effects will help keep the viewer's attention. Use a professional sound studio when recording a narrator. The studio's engineer should be able to help you select music to complement and enhance the presentation. You'll be surprised to find how inexpensive it is to add music — especially considering its impact on your program.

38

Five more tips for better slide shows

Here are five more tips to help make sure your slide presentations are engaging and to the point:

1. *Make the graphics consistent.* When discussing title slides with whomever is doing the graphics, emphasize the importance of maintaining the same "look" throughout the program. Any change in the visual rhythm of the presentation will cause the viewer to stop and ask why.

2. *Watch out for visual overload.* A common mistake is to cram too much information onto a single slide. Use the six-by-six rule: no more than six lines per visual and no more than six words per line.

3. *Use photography effectively.* Hiring a professional photographer is a good idea, not a frill. You'll ensure consistency and quality for the finished product. Provide your photographer with a written "shot list" of all the specific photos you want taken. Let the photographer know you'd like to see a variety of camera lenses and angles used. The more information you give your photographer, the better results you'll get.

4. *Use a variety of visuals.* Keep presentations moving by blending different types of slides — graphic and photographic, for instance — or

combining both on a single slide for some fascinating effects. Use appropriate graphic slides to emphasize, not repeat, the narrator's words. Dramatic, storytelling photography can show the viewer exactly what the script says. By using a variety of visuals, you'll keep the program from slowing down and becoming predictable.

5. *Distribute the show cost-effectively.* While you may have produced the show in slide format, that may not be the most economical way to distribute it. Do the trainees have easy access to sound/slide projectors? How about video playback equipment? The answers to those questions will point to the most appropriate format for distribution. If participants have access to both and depending on the number of slides and number of copies required — videotape transfer may be cheaper. If you decide to go with slides and need to send out duplicate sets, put a few drops of strong glue on the tray ring or dust cover. This tactic will ensure that the slides stay together during use or shipping.

39

Showing slides featuring trainees builds rapport

Before her training sessions, Pam Pinney, an industrial engineer with Whirlpool Corp. in Marion, OH, takes slide photographs of trainees-to-be operating in their work environments. She then flashes these individual snapshots on the screen for about five seconds at the opening of her session or upon return from breaks.

Pinney says taking the photos not only helps her become more familiar with the type of work participants do, establishing stronger rapport, it also gives trainees some additional recognition and challenges them to match names with faces as the slides pop up on screen.

■ To keeps things light, Nick Baldwin, a program specialist for the Department of Education in Florida, sometimes includes off-the-wall photographs in his slide shows, showing something that a trainee has done on the job that's humorous, but not embarrassing.

S lides of workplace situations are the basis of a review process used by Ken Moldenhauer, training director at Excel Corp., Dodge City, KS. Instead of going over a lot of notes and having students repeat what they wrote during lectures, he shows slides illustrating concepts that were covered in the session.

Participants volunteer information about things they see in the picture that pertain to their new skills and knowledge. Instead of merely repeating what they were told, they get a chance to apply class materials much as they will when they return to the workplace.

40

Slides of workplace situations work well as review tools

Chapter Four Tips: Videotapes

41

Videotapes show trainee progress

Many train-the-trainer courses call for attendees' presentations to be videotaped for review later in class. In an effort to help lower materials costs and to make sure trainees have something to take away with them, Nancy Nicolazzo, president of Integrity Seminars, Durham, NH, asks participants to bring a blank video tape with them to class. That way, the tapes can be used for recording and review during class, and participants can then take the tapes with them to see how far they've come after they've become experienced.

Darcy Karlen of Gilbert/Robinson Co. recommends using clips from popular films — renting videotapes and utilizing selected portions with a VCR — to illustrate training points. To model self-confidence, for example, you could use clips from Robert Redford's triumphant scenes in *The Natural*. Or perhaps the warehouse scene in *Beverly Hills Cop*, where Eddie Murphy uses quick thinking to pass himself off as a security inspector, can be used to set the stage for talking about effective security measures.

■ Terri Weeks, a selling symposium trainer, uses a short clip from the movie *Top Gun* to introduce her symposium. Because the purpose of the selling symposium is to encourage participants to be the best they can be and to create a sense of teamwork — "We're in this together!" — she finds the film clip, which emphasizes teamwork, helps participants relate to her content from the beginning.

42

Silver screen snippets set the stage for training

43

Video camera broadcasts flip charts for large groups

High tech meets low tech when a flip chart is the best medium to deliver your message in a classroom but the class is too large for paper and markers to be effective. Consider using a video camera to broadcast close-up pictures of your work on the flip chart through a series of video monitors or a large-screen projection system. This also is an effective technique for showing any props or small items you may want to use to illustrate pertinent points. Adding a video camera to your list of audiovisual equipment for a presentation is especially cost-effective if you already plan to use monitors or large-screen projectors for films, videos, or other presentation systems. This is a technique that can be used for groups of 1,000 or more.

The next time you have a film or video to show during class, consider designing the class so the film is shown just before lunch. During your preliminary remarks give an overview of what to expect, including the topic and length of the film, and emphasize two or three questions that you would like participants to consider.

But once the film is shown, break for lunch. That not only gives people time to consider the film, it provides continuity. After lunch, use the discussion questions to review, reinforce, and generate participant comments.

As an alternative, preview your video to see if it can be broken into two segments. You can play part one before lunch, and after lunch hold a discussion revolving around part one. Then show part two before the afternoon break and hold another discussion after that break.

44

The best time to show videos is before a break

45

Video camera hook-up brings reality into the classroom

Sometimes reality is the best teacher, and creative use of audiovisual equipment can bring real situations to the classroom. For example, Deanne Gesdahl, instructor/designer for Mercy General Hospital in Sacramento, uses a video camera hook-up while teaching a laser surgery seminar.

A camera in the operating room televises the technical process live to participants in the hospital auditorium. A speaker in the operating room allows two-way conversation with participants and the physicians performing the procedure. This technique has a powerful impact on the participants' interest, motivation, and enthusiasm.

46

'What We Do' videos showcase different departments

To bring the organization closer together and facilitate communication companywide, Paula Anderson-Findley, training and development manager for Sunburst Farms, says her company of 135 created a "university."

Each department put together a 30-minute "What We Do" presentation. All employees go through the training program to discover what goes on in other departments and to improve internal customer service.

■ Trainer Penni Helsene of Citizens Insurance Co. of America, Howell, MI, also uses videos to educate employees about departments within her organization.

Helsene enlisted managers of various work areas to act as consultants on the videos. She wrote scripts and hired a video company to edit and film them. The videos each have a manager describing his or her department and key employees. The audio is overlaid with pictures of employees and text showing titles and key functions.

The videos not only minimize boredom in presentations, they involved managers in the training function. The videos are cost-effective, too. They are used for new-employee orientations and at branch locations.

47

Producing videos makes for an engaging review

Not only is everyone a star when participants make videos for the rest of the class, but they learn in two ways. They review information through their involvement in preparing the topic for presentation and also by watching videos others have made.

One-minute videos can be hilarious takeoffs on commercials while reinforcing a point and breaking the routine of a classroom training program. Once you have finished a major section of the program, allow groups of participants three to five minutes to come up with a one-minute commercial, rap performance, skit, or song that reviews the material. Better yet, have them think about possibilities over a break or lunch and then come back to brainstorm and produce their videotape. A breakout room can be set up as a studio to videotape what they have created. Then the one-minute "infomercials" can be used later as an energizing and humorous review.

'Blooper' video shows trainees practice makes perfect

When teaching a course to fellow presenters on video desensitization, Susan Thornton, a nurse and trainer at the Tucson Medical Center, shows the class a video of her rehearsing the presentation — flaws and all. She says the "blooper" film warms people to the idea of being on camera and lets them see they don't have to be perfect the first time. The film also helps make several other points:

• Novice presenters can see that even experienced speakers extensively rehearse their presentations.

• Attendees see they can rehearse in a private, quiet location and deal with their mistakes without an audience.

• A dress rehearsal is possible, allowing the presenter to try on clothing that would be suitable for on-camera work.

• By watching themselves on tape, novice presenters begin to be more critical of their presentation techniques.

• Errors in dress, movement, and stage presence are recorded in a visual format the presenter can

49

Rid your VCR of unwanted noise

What R.L. Vanover, a regional training manager for Lederle Laboratories in Cincinnati, thought might be an expensive problem with unwanted TV noise when reviewing videotaped role-plays in his training sessions was solved with a couple of inexpensive items.

In those training sessions, he was videotaping role-play sessions using a standard half-inch, VCR and TV connected with coaxial cable. But the setup had one drawback — unwanted sound. During the review portions of the role-plays, Vanover would pause the video to comment on the presentations. When his comments exceeded a few minutes, however, the VCR was programmed to "fade to black" and the loud refrain of a TV tuned to a nonexistent station followed.

He thought the problem could only be corrected by replacing the TV with a video monitor. But he discovered by simply inserting RCA plugs into the "video out" and "audio out" jacks on the VCR, there was no longer any unwanted noise when the TV went to black.

Vanover says the inexpensive remedy works even if the plugs are not connected to anything on the other end.

Contrary to popular opinion, training videos don't have to be expensive to produce or a complete bore to watch. By turning the project over to employees, you can create a vast amount of excited ownership in the video and save your company money to boot.

Need evidence? Just ask the truckers at Motor Cargo. There, a team of eight truckers and a dispatcher produced an entertaining, informative video for co-workers on a tedious topic: How to catch billing errors to inaccurate shipment descriptions.

"Employees know better than anyone what work habits need sprucing up and viewed producing the video as a welcome break from the routine," says Kevin Avery, employee-involvement coordinator for Motor Cargo, Salt Lake City.

The team of truckers donated 400 work hours to produce, direct, and act out scenes of billing mistakes. Motor Cargo paid $1,500 for the editing. The payoff: In a test study, $16.48 per bill was saved because of more accurate descriptions, and checking took only five minutes.

50

Employee-made videos tap those closest to a topic

51

Playing music through a VCR produces high-quality sound

Many trainers who use music in the classroom find it's difficult, inconvenient, or expensive to have a high-quality cassette player in the training room. And smaller, more portable players are incapable of producing good sound.

Jim Van Dyk, corporate training supervisor with Triad Systems Corp., Livermore, CA, solves this problem by transferring music to video cassette format, with the "snow" blacked out (a service available from many high-tech video outlets or internal media departments, according to Van Dyk). The ever-present video equipment in a classroom produces high-quality audio, and the result is an instantly available audio player.

To break up the tedium of lecture, Dottie Knisely, a sales trainer with Microsoft in Redmond, WA, asks managers and other subject matter experts in her company to put together 15- to 30-minute videotape presentations on some aspect of their expertise pertaining to class content. She then plays the tapes at predetermined intervals to break preoccupation.

"I get tired of hearing my own voice, and I know trainees can too," Knisely says. "And when the subject matter experts can't come to class in person, it's a good way for trainees to hear information straight from the horse's mouth." The video presentations typically cover subjects such as computer hardware repair, handling product returns, improving phone skills, and communication skills.

52

Subject-matter videos break tedium of lecture

53

Worksheets keep trainees awake during videos

When the lights go down and a video goes up on the screen, some participants may be tempted to take a little mental break. To combat this tendency David Allgood, manager of training for Egghead Software, developed "video discussion sheets" for all of his classroom videos.

Simple fill-in-the-blank questions help participants pay closer attention to a video and retain more information, he says. Allgood has participants read the questions before the video and then fill in the blanks as they watch. After the video, he uses the sheets for discussion and emphasizes their importance as a review tool.

M ark Long, program development supervisor for Mazzio Corp., uses a video camcorder to interview training participants prior to his sessions and during breaks about their expectations, individual needs, common problems, breakthrough ideas, etc.

At the end of his four-day training session he shows an edited 10-15 minute video presentation as a closing that helps build post-class networking.

54

Videotaped interviews with trainees foster post-class networking

55

Showing employees having fun is effective orientation

A lighthearted, "homemade" video is a critical part of orientation training for TRW Target Marketing in Orange, CA, says Ann Stenger, technical training consultant. The video shows current associates having fun at work and gets new employees laughing while introducing them to their new coworkers.

Stegner roamed the halls of the company to capture on film as many workers as possible. She asked them to sing a favorite song, perform a favorite dance, or do a favorite impression, etc.

Chapter Five Tips: Various Audiovisuals

56

Personal tape-recorded reviews aid retention

Having participants create their own personal review on a cassette tape — instead of just writing key learning points — helps pull participants into the review session and aids retention, according to Terry Paulson, president of Paulson & Associates, Agoura Hills, CA.

Throughout class, participants make lists of key thoughts or quotes that help remind them of meaningful course content. They then record the items on a cassette, five to 10 items at a time. After each set of items attendees add a short musical interlude of their favorite music. The musical interludes allow for reflection on the preceding "keepers." The process is repeated until all material is covered. The result is a personal review tape that won't get tossed aside as written notes sometimes do.

Most people have an audiocassette player in their office, car, home, or all three. So instead of sending the usual invitation to — or reminder of — training by memo or letter, Mike Redwood, a trainer with the Cryovac Division of W.R. Grace of Canada Ltd., Mississauga, Ontario, sends a tape of his voice.

The message can be easily recorded in your office, he says, and you can add your choice of music for background. With a dual cassette player and a minimal expense for blank tapes, you can then make the desired number of copies.

The tape, Redwood says, provides a great medium for letting participants know when and where the session will take place, what it has in store for them, and serves as an informal introduction of the trainer to those people who may not be familiar with him or her.

57

Cassette tapes make unique training invites

58

Use sounds to illustrate key points

When a presentation for a large group makes it difficult for all participants to see visuals, such as overheads, flip charts, and so on, sound can be used effectively to illustrate a point. Pamela Jamar, a senior consultant for Software Usability Specialists, Minneapolis, suggests using a bell, buzzer, or chime. The sounds give people notice — without making them look — that an important point is coming up, she is changing direction, or that they should redirect their attention. She says using a change from visuals to sounds as a good way to wake people up.

59

One cartoon is worth a thousand words

If one picture is worth a thousand words, and pictures can enhance your training message, then it's probably worth selecting key course concepts you want to reinforce in a program and having cartoons drawn to illustrate them. Try to illustrate the entire concept without any words at all. The final product can be used as an overhead, as posters for display in class, and as handouts for participants. The illustrations are especially useful in classes where several languages may be spoken.

Your area university art department or vocational school graphic design program may be a good starting point to locate budding artists. Or you might find people in your company who do cartooning as an avocation.

■ The comic strip format is great at conveying procedural information, says Charlotte Hughes, training specialist with Dreyfus Inc. The steps to take in processing a request for information or a request for a refund, for example, become much more lively when presented in a paneled format, however simple the illustration. It is fun for participants to read, can be entertaining as well as educational, and is more likely to be recalled.

60

Tap artistically inclined employees for classroom graphics

Trainers who consider themselves artistically handicapped can be at a disadvantage when it comes to decorating a training room and making it a little more hospitable. Diane Maze solves the problem by asking employees in the company know to be artistically inclined to design posters and flip chart sketches. Everyone benefits, Maze says. Trainees love the additions to the classroom, the trainer is able to enliven the learning environment, and the resident artists gain recognition around the company.

Cartoons or magazine illustrations can be enlarged to provide inexpensive, thought-provoking posters for your training classroom. Sherrie Spilde, training specialist with the state of South Dakota personnel bureau, has a simple technique.

She makes a photocopy of the art on an overhead transparency, projects it onto a flip chart or poster board, and traces the enlarged image with markers. In addition, the images are easily stored or transported in cardboard tubes.

61

Poster-size cartoons make thought-provoking visuals

62

Sound system tips ensure entire audience hears

When using a sound system, check it out yourself in advance. Here are a few tips for using a sound system:

• Use a sound system if participants will be seated more than 30 feet away from the presenter, if the group is larger than 25 people, or if the environment is noisy, a sound system should be used.

• Remember that more volume will be needed as the room fills.

• You need to become comfortable with microphones. A handheld mike works well held 6 to 10 inches from your mouth and at roughly a 45 degree angle. Too far away and your audience will not be able to hear you, too close and every voice defect can be heard, every sliced "s" and popped "p."

If a room is very noisy, hold the microphone beneath your chin to increase voice volume without creating a blast effect. If you plan to use a lot of visual aids or to move around the room much, consider a clip-on microphone.

• Many presenters today use wireless microphones, which cost between $125 and $4,000. Remember to also purchase an AC line filter. One advantage of a wireless mike is that it enables you to easily tape record your presentations.

Most FM transmitters for wireless mikes have a tape recorder jack, which gives you much better sound quality than if you simply set up a tape recorder with a separate microphone. If you decide to use this approach you might also want to consider a voice-activated tape recorder. Jacking this into your microphone transmitter it allows you to start and stop your tape recorder by simply using the microphone on-off switch. When the switch is in the off position the voice-activated tape recorder receives no signal and stops, so you can save tape and editing time when you are doing activities you don't need to record.

• Make sure you check for feedback. Be sure that your test includes all of the components in the places they are going to be used. If you are going to walk around in your presentation, make certain you walk around as you test the mike. There is nothing worse than to begin to move about the room and find that your microphone squeals as you move too near a speaker.

• And if you are going to be using a microphone outside, put a wind screen on the microphone, a small sponge-like cover for the microphone head.

63

Lighting tips focus attention on the speaker

Take a critical look at the lights every time you use a new meeting room. Never assume the lighting in the room is the lighting that has to be used. Better lighting is almost always available, and it is particularly important to focus on the area where you will stand. For example, lights can be unscrewed from the ceiling and replaced with gooseneck spotlights. Track lighting can be rearranged to focus more light on the presenter.

It is distracting for participants to be unable to see the facial expressions of the presenter. And there are few things more uncomfortable for you than to feel confined to a small three-foot area, fearing that if you step out your face will be nothing more than a silhouette.

Attorney Robert Jacobson of Palmatier & Sjoquist, a Minneapolis law firm specializing in patent, trademark, and copyright matters says that in an educational setting, there is some leeway in copyright laws about the need to obtain permission for using copyrighted material. Under the "fair use" clause of those laws, four factors come into legal play in determining whether the use of copyrighted material is a fair use:

The purpose and character of the use, including whether such use is of a commercial nature or is for a nonprofit educational purpose.

• The nature of the copyrighted work.

• The amount and length of the portion used in relation to the copyrighted work as a whole.

• The effect of the use upon the potential market for and the value of the copyrighted work.

So based on the fair use clause, it may or may not be necessary to obtain permission to use copyrighted work. But Jacobson recommends consulting an attorney if you are considering using such material in your program. Getting legal advice first might save you the need to obtain permission or the cost of an infringement suit, he says.

64

Using copyrighted work? Keep these guidelines in mind

65

Five tips on when to use visual aids

Presentation software has put quality visuals at the fingertips of every computer user. But it has also raised many questions for presenters, the first of which usually is, when should visual aids be used? These five tips from 3M Visual Systems Division, St. Paul, MN, spell out the wheres and whys of visual aid use:

1. Use visuals to get the audience's attention. Kick off the meeting with a real attention grabber, such as a cartoon or famous quotation.

2. Use visuals to emphasize key points of the presentation. Ask yourself, what are the three to five issues participants should remember when they leave?

3. Use visuals to present statistical data in an easy-to-read format, such as a bar graph or pie chart.

4. Use visuals to compare data, such as sales against forecasts or trends over time.

5. Use visuals to show items too big or too small to display at a meeting, such as a large machine or computer memory chip.

For more information on effective use of visuals, contact the 3M Visual Systems Division, 3M Center, Building A145-5N-02, Austin, TX, 78726-9000, (800) 328-1371.

Interesting illustrations and graphics for flip charts are as close as your office bookshelves, says Tim Tomczyk of Greyhound Leisure Services. In preparing for a course, Tomczyk, with an eye for relevant art, checks his books related to the course topic.

He enlarges the graphic on a photocopier, colors it with markers, and cuts it out. Once it's pasted in position on the flip-chart page and the text is added in a complementary color, he has a colorful and interesting poster.

66

Sources for visuals are as close as your bookshelf

67

Ideas for updating old visuals

If you're tired of your old visual aids but don't have the time to develop new ones, try these ideas:

Try tackling just one part at a time — the Swiss cheese approach. Instead of facing the enormous task of developing all new visuals, start by choosing one visual that you'd most like to improve. Keep that in mind as you teach the course. Ask yourself if there's a way the visual could be used more effectively. Could the colors be changed? Could the visualization of the concept be improved? Could your copy be "tightened up" and better targeted to your audience? Jot down any brainstorms you have along the edge of a piece of paper that has a copy of the visual on it. And make sure you go through a dry run with the new visual before unveiling it in your training program. Your trial audience may be able to make even further suggestions for improvement. After a short period of time you will undoubtedly come up with ideas for ways the visual could be used more effectively.

Once you've taken the time to put that idea into action, pick another and take the same approach. As you see the impact the new visuals have, it will accelerate the process of replacing the others.

Storyboarding is a technique used by professional scriptwriters and advertising writers to depict consecutive key changes of scene in a planned film.

It's a tactic that can be adapted to training visuals. Jim Davis, senior training and materials specialist with Hallmark Cards, leaves the left third of his pages blank as he writes training programs so he can make rough sketches and notes alongside the script.

Storyboard technique adapts well to developing training programs

Example of the Storyboarding Technique

Start new section on goal-setting.

Play the song, "To Dream the Impossible Dream"

Give the definition of goal-setting.

Show an overhead of a person reaching for a cloud.

69

Magnets effective learning tools for building models

If the chalkboard or whiteboard in your training room has a metal base, using magnets can turn it into a more effective visual aid. Magnets are especially effective with presentations that build models for participants as each part of a process can be depicted separately, then added to.

Illustrating a model with magnets is simple and inexpensive. Foam board and adhesive-backed magnetic strips are available at most office supply or craft stores. Write the key words or draw the principles of your process on the foam board, cut arrows from the foam board that point to the upcoming items, and attach magnetic strips to each piece. As you make individual points, place the word or picture that corresponds on the board, using the arrows to signify transitions between ideas. The technique keeps the ideas "in front" of participants and helps build the presentation effectively.

70

Idea file handy for spicing up visuals

Need to design a catchy cover for a guide or transparency and you're out of ideas? Drained of every ounce of creativity? An idea file can be a great way to get the creative wheels moving when you feel stumped, says Michelle Tomczyk, project manager of instructional design at Automatic Data Processing Inc. in Parsippany, NJ.

She suggests keeping a file that includes photocopies of interesting book, manual, and magazine covers, training literature that catches your eye, and direct marketing that uses a fresh design approach. The evening news program on TV — where background graphics are used to reinforce the spoken word — is also a great source for ideas. Tomczyk watches for interesting ways that print type, layout, color, and shading are used to enhance the message, makes a note of the techniques, and puts them in her idea file.

71

A simple solution for when there isn't a dimmer switch

Many rooms used for training are not designed with the express intent of holding training sessions. For instance, some rooms don't have lights with all-important dimmer switches or rheostats. Rickey Peck, a training assistant at Kerr-McGee Coal Corp., has found a simple way around that problem. Instead of totally darkening a room to show films and slides, Peck places a lamp with a small wattage bulb in a rear corner of the room to provide light when main lights are out. The light can be turned on at the beginning of the program and remain on throughout.

For trainers who are "on the go" and do not have the luxury of a home base of operations or knowing what materials will be waiting for them to use at the training site, Ashley Fields recommends a "survival box" of materials. All the items listed will fit into a copier paper-sized box or portable file box:

Supply Checklist

Name tags	1 box of 50
Name "tent cards"	25-30
Transparency sheets	12
Sheet protectors (transparencies)	12
Dry erasure markers	1 box of 4
Flip-chart markers	1 box of 6
Black permanent markers	6
Pencils and pens	12 of each
Highlighters	12 (scented)
Masking tape	2 rolls
Overhead pens	1 box
Paper clips	1 box
Binder clips (to mount flip-chart sheets on walls with nails)	25-30
Post-it Notes	3 pads
Business cards	25-30
File folders (stuffed with 10 sheets of assorted, similar colored paper)	6
Scissors	1 each
Ruler	1 each
Note pad	1 each

73

Paper 'clam' holds tape loops till they're needed

Tearing pieces of tape during sessions to post materials on classroom walls can be a hassle, says Genie Wilson, a trainer and facilitator at Texas Instruments, Richardson, TX. And sometimes it's just not possible to be there early enough to do that sort of prep work before a session begins. Wilson recommends getting two paper plates before your presentation, preparing as many tape loops as you expect to need, and sticking them loosely to the surface of one of the plates. Then place the other plate face-down as a cover — so the plates take the shape of a clam — and tape it in place on one end. This keeps the tape loops from getting stuck to one another or to the lid. Just remove the "lid" to use.

S ometimes props make detailed processes easier to visualize, says Cindy Turbowan inside sales trainer at Microsoft, Redmond, WA. For example, when teaching employees the steps involved in the product return process, she might begin by holding up an example of the product to be returned, as she explains the defect or reason for return. Next, a telephone, as she talks about the customer's call to the service center. A toy shipping van might represent the return shipping of the product, and a company warehouse worker's hat might symbolize the arrival of the product back at the company shipping center.

People are able to picture the objects shown when trying to review the steps in their minds, she says. The system works for just about any process, and is epecially effective for tactile learners.

74

Props help trainees remember processes

75

Music sets a mood to learn by

Music can set a mood to learn by. Consider compiling a varied selection of tapes participants can listen to during breaks. You don't need anything too formal —just tapes, an uncomplicated and reliable tape player, and easy access so participants can handle the details themselves. Be sure you allow people the freedom to indicate when the music becomes bothersome.

Plan also to use the music during individual or group activities. Music such as Vivaldi's *Four Seasons* or Steven Halpern's *Spectrum Suite* can help create an environment conducive to learning.

■ Music can also be used to promote and reinforce course content. Harriet Reichman of Citibank sent her participants a tape before classes began with an upbeat introduction to her program. Then she had a professional singer record (with an upbeat popular tune) selected key ideas covered in training to follow-up and reinforce the training. The follow-up tape was sent to participants about a week after the training program ended.

Chapter Six Tips:
Audiovisuals as Learning Tools

Tips continued on p. 96

Participants share their thoughts on a large roll of paper, such as butcher paper or newsprint, taped across a wall in classes taught by Dennis Link, operations training coordinator for Amoco Oil in Texas City, TX. He divides the paper into three sections labeled: Issues/Concerns, Main Ideas, and Thoughts/Feelings. Link encourages people to express themselves on the wall. The process helps others to learn and to remember key points on the wall. He says it is a tremendous way to keep trainees involved in training.

76

Open-canvas wall lets trainees express themselves

77

Puzzle exercise emphasizes learning objectives

Often when course objectives are printed and handed out with other course materials, their importance is lost. Lori Smith, staff development coordinator for South Community Hospital, suggests emphasizing the objectives of a session through this exercise:

She prints each objective in large type over an illustration, and then cuts the illustrations into enough puzzles to give one piece to each participant. When it's time to discuss course objectives, she tells the class the objectives will be discovered as the puzzles are put together. The task is to find out where they "fit" with the objectives.

She then gives each small group (formed by joining their puzzle pieces together) time to discuss what each member would like to get from its objective. She has a leader from each group state the objectives to the class and share a few of the highlights of their discussions.

Smith says the activity acts as a quick survey for the trainer to see where everyone is on the objectives. She says it has been especially effective with ongoing programs, such as orientation and safety training.

A collage created by trainees on a classroom wall that depicts major learning points acts as a review for students of Cathy Cady, manager of education for Southwestern Bell.

Cady cuts pictures from catalogs and magazines and brings them to class. Throughout a week-long class, she has students choose pictures to paste into the collage that reinforce lessons taught. As the collage grows it becomes a reminder of the major learning points. At the end of the class, the completed collage is given to a student as a reward.

■ Inviting participants to create a collage relating to class concepts is a relaxing, thoughtful way to conclude class, says Amy Gonnella, director of public education for the American Cancer Society.

At the end of her training programs, she gives participants time to put together a collage of goals, leadership skills, or visions that are personal or career oriented. She provides paper, colored pens, glitter, magazines, glue, and background music. The project gets people involved because their goals have become concrete and they have a personal reminder to take with them from the training.

78

Collages foster creativity, double as review tools

79

Target pinpoints whether a session hits the mark

How "on target" are your presentations? Dara Lowry, Riverbluff Council of Girl Scouts, uses a traditional, circular target for people to indicate their satisfaction with her training. Lowry has participants mark a target as they leave the seminar or at the end of each day to give her a feeling for whether the course met their needs.

Outside the target, around the first ring, reads, "Way out in left field." Inside the first of three rings is printed, "Not what I needed." The second ring says, "Almost but not quite." The bull's-eye is labeled, "Hits the mark." This graphic may be blown up and photocopied to see how "on target" your training is.

How on Target Was This Training Presentation?

Not what I needed

Almost but not quite

Hits the mark

Way out in left field

Creating a logo for training can help reinforce your program's themes — before, during, and after a program, says Katherine Brown, chief of staff and faculty development for the U.S. Army Chaplain Center in Ft. Monmouth, NJ.

Brown developed a logo for a training retreat with the theme, "Working as a Team: Handling Change Constructively." The symbol was printed on name tags and memo pads and mailed in a packet to each participant in advance. The logo helps familiarize participants with the theme before class begins, emphasizes the message during the workshop, and reinforces seminar content back on the job.

80

Creating a logo reinforces a training theme

81

Unique visuals help trainees remember learning points

This memory device used by T.J. Titcomb, director of training at Family Service in Lancaster, PA, in presentation skills or train-the-trainer seminars adds interest and helps novice presenters break bad habits.

Titcomb takes each basic training skill or hint, pares it down to a catchy "law," and then matches it with appropriate visual clues. As an option, she also plays an appropriate "soundtrack" with a cassette recorder. Some sample theorems she uses include:

• *Lloyd's Hypothesis: Begin and end on time...*

Present a transparency featuring the classic photo of screen star Harold Lloyd dangling from a huge clockface over a city street many stories below.

• *Duke of Windsor Dilemma: Abdicate the Power Role...*

Present a picture of the Duke and Duchess of Windsor and give the explanation: "In 1936 King Edward VIII gave up his throne for the woman he loved. If you love training and want real learning to happen, you also have to step out of the power role. Your focus should always be on participatory learning activities rather than transmittal, lecture-based programs."

• *Lee's Premise: Never apologize...*

Present a picture of an old 45 rpm record while playing Brenda Lee's hit, *I'm Sorry*. Explain that trainers should not draw attention to perceived deficiencies in their presentations with unnecessary apologizing. If a real problem exists, recognize it and state what is being done to correct it.

• *White Rabbit Rule: Never look at your watch...*

Present an illustration from *Alice in Wonderland* of the White Rabbit holding its watch. Solicit participants' input as to what happens to learners when the trainer looks at her watch. Stress that the learning process is interrupted by this activity. (Hint: A small clock can be hidden from participants' view on the podium or near the overhead projector.)

82

Visuals lay the groundwork for day's learning

A "question of the day" followed by a series of "guiding footsteps" lays the groundwork for the day's agenda in training classes conducted by Linda Simon, owner of Fit to Live Training and Development, Minneapolis.

A poster with a "question of the day" related to the main course topic greets participants as they enter Simon's classroom. Footsteps posted on the floor, walls, or ceiling — each detailing a step of the process or a key idea relating to the topic — lead the trainees throughout the room.

In addition to piquing curiosity and introducing the subject matter, Simon uses the footsteps as an energizer and review technique. Anytime participants look like they could use some time out of their chairs she has them get up and literally "walk" through the points she's delivered, step-by-step.

Simon cuts the footsteps from construction paper and laminates them to make them more durable.

A "duty board" is helpful for keeping group projects on track, says Jackie Runnion, an instructor at Bellingham Vo-Tech Institute, Bellingham, WA. The board — any bulletin board or wall space will work — is divided into three columns: To Do, Doing, and Done.

Each step of the project is written on a separate Post-it Note and posted on the "To Do" section of the board. Each task is assigned to a group member, whose name may then be added to the note. As each person starts a task, the appropriate note is moved into the "Doing" column, and finally into the "Done" column as the task is completed.

At the bottom of each column is an area marked "Questions/Problems." Any group member may place a note there, for consideration by the instructor or the entire group.

The duty board process keeps the group from overlooking portions of the project, tells members whom to consult about a given part of the project, and gives participants a way to keep tabs on the overall progress of the project.

83

Duty board keeps group projects on track

84

Drawings help trainees remember by association

Researchers have determined the average person can only hold seven pieces of information in short-term memory, plus or minus two pieces. So creating a "windowpane flip chart" can help people visualize and better remember a concept. Here's how it works:

Divide a flip-chart page into six windowpanes, and in each square place a picture that represents the idea you're trying to get across. Use short words if needed, but try for pictures that don't rely on verbiage to drive home the point.

Here are six examples using simple line drawings:

1. A computer illustration shows the concern trainers have with the rapid increase of technology on the job, and the challenge to train people based on that rapid growth.

2. A pair of ballet slippers or dancing shoes represents the concern about balancing the needs of the individual versus the needs of the organization in designing and delivering training.

3. A book with an arrow emerging from it and pointing to a computer disk illustrates an explosion of information and the fact that it's no longer possible to teach everything.

4. A star hatching from an egg illustrates the training's concern

with "hatching," nurturing, and developing training superstars.

5. A prison with the symbol for "No" or "Don't" represents the desire to retain good employees without having them feel like prisoners.

6. An exit sign represents how companies downsize when it's necessary while maintaining a concern and responsibility to the community and employees who leave the organization.

Try matching what was described with the graphic below. Then cover the text and attempt to remember what the pictures stand for. You should be able to remember most of the concepts just from this brief exposure.

85

Review tree lets trainees keep their questions anonymous

Rick Kitchen, a training supervisor with C.U.C. International in Dayton, OH, uses a "review tree" to ensure he addresses all pertinent questions before class ends — particularly questions from trainees reluctant to speak up for fear of asking "silly" questions.

He first draws a tree with multiple branches on a sheet of flip-chart paper. The session's main topic forms the "trunk" of the tree, and individual categories of the topic make up the branches (see graphic on opposite page). In a session, for instance, on "voice presentation skills" as the main topic, the branch categories would be speech inflection, tone, pacing, and so on.

The branches — which remain covered by slips of paper until the category is addressed in class — are drawn without leaves. During breaks or after a session, Kitchen asks trainees with questions that haven't yet been answered to write the question on a large Post-it Note, then stick the note/leaf on the branch subject it best fits. "We ask them to post it even if the same question has already been posted," Kitchen says, "to help gauge how much more review is needed." It's also an effective way for shy or reluctant participants to get questions answered, he says.

Questions that don't fit categories are placed on the "ground" — the bottom of the flip-chart page — as fallen leaves, and are also addressed.

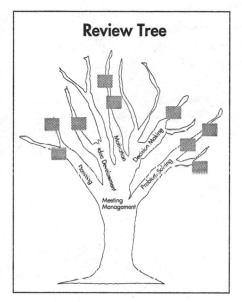

Review Tree

Pruning
Idea Development
Motivation
Decision Making
Problem Solving
Meeting Management

86

Cartoon caption-writing activity adds levity

Topical cartoons — minus their captions — add levity to dry subject matter during Mark Poore's software training programs. Poore, a training coordinator at Precision Computer Systems Inc. in Sioux Falls, SD, displays cartoons relating to computers on an overhead projector. Before revealing their captions he solicits ideas for substitute captions from the group. These are often funnier than the original punchlines, he says. Try guessing on the cartoon below yourself.

The technique has broad applications, Poore says, as cartoons can be found relating to most topics.

Answer: "See you later, I've got training class."

87

Caricature drawing exercise energizes class

P articipants work in teams to "illustrate" the attributes and characteristics of the ideal service provider in customer service courses taught by Gerri Hura, director of training at Vista Host, Houston, TX. She suggests this exercise as a fun energizer and as a way to facilitate team-building.

Hura first has the class brainstorm on the qualities and attributes of the "ideal" service provider and lists their responses on a flip-chart page. Some examples include smiles, professional dress, good listening skills, etc. She then breaks the group into small groups and assigns each group a part of the body: head, torso, and legs (for three small groups) or head and body (two groups). She gives each group a flipchart page and markers and tells them to draw the body part to emulate the ideal service provider, using the attributes or qualities they've described to create caricatures, accentuating body parts to illustrate characteristics. For example, big ears indicate a good listener or tennis shoes show the person is quick on their feet.

After 10 minutes she has each group member "autograph" the picture and tape the drawings to a wall to make a complete body.

88

Letting trainees act as instructor breaks monotony

An endless string of slides detailing work processes can be boring and taxing for trainees to sit through. To break the monotony, Lisa Murphy, a trainer of with Hoechst Celanese, Textile Fibers Group, Rock Hill, SC, lets trainees act as the "instructors" during reviews.

In her technical training classes, she uses flow charts supplemented with slides to illustrate steps of a process. After going through the flowchart and completing a question-answer period, Murphy resets the slides back to the beginning and asks a student to be the instructor. He takes that attendee's name tag and seat.

The student then proceeds to reteach the task and explain each slide. Murphy becomes the most inquisitive student in the class. Any question concerning material that has already been discussed is "fair game" for her to ask. She says it adds humor to the class when she exclaims with a confused look on here face, "I don't understand." Her request for a better explanation encourages the student "teachers" to draw on the board, write definitions on a flip chart, and even call on their classmates for help.

Barbara Taraskiewicz, training coordinator for Kalamazoo Valley Community College, uses the common brainteaser pictures of the chalice/profiles and the old woman/young woman for an introductory exercise that emphasizes the importance of focus and perception.

Rather than challenging participants to tell her what they see, she helps participants see both images and then points out that what they see first depends upon their focus of attention. She makes the point that we are prisoners of our perceptions and tend to only see what we're focused on, and that once we're focused, it becomes difficult to see other things that have been there all along.

Brainteaser helps trainees see from another perspective

90

Doodling helps trainees unleash creativity

Time-management workshops are often uncomfortable for participants because no one wants to admit they have a problem managing time — especially managers or first-line supervisors. Rosalie Webster, training specialist at Crum & Forster Corp., maintains that asking each person to write or talk to the group about their problems is a one-way road to boredom.

She avoids that trap with a step-by-step exercise that breaks the ice and makes a valuable point regarding time management:

• Spaciously post flip-chart sheets around the room, one for each person. Use tables, walls, the floor.

• Give each person markers.

• Introduce the workshop by stressing that time problems have no restrictions; they are as much a part of our daily personal lives as they are in the workplace.

• Give participants a moment to think about a "real" time-management problem.

• Ask them to draw or doodle their time-management problem on a piece of flip-chart paper.

Webster says participants unleash their creativity as they doodle. When everyone is finished, she has each person explain his or her art.

P oster-sized photos of everyday items stimulate discussion on the importance of perspective, says Don Allison, national sales trainer for Hallmark Cards, Kansas City. Calling his activity the "Power of Imagination," Allison uses photo blow-ups to illustrate how using unique ideas or techniques can make basic information more interesting.

The photos are taken at angles that make identifying the objects difficult. The objects are "blown up" larger than life, or show a close-up of a small part of the subject. Allison's blow-ups include a donut, a turtle shell, an alligator snout, a piece of grass sticking up out of the snow, a portion of hair on the back of someone's head, and the closure at the top of a purse.

He hangs 20 of the posters in the meeting room and asks participants to try to describe what each poster is on an entry form. The reward for getting the most correct answers is a dinner for two.

Using strange photos underscores need for perspective

92

**'Dotmocracy'
determines
learning
needs
of group**

Max Crotean a management development consultant at Ontario Hydro, Toronto, Ontario, uses "dotmocracy" to determine learning needs in his classes.

Crotean lists the topics to be covered during his course on sheets of flip-chart paper and posts them in the classroom. At the beginning of class, he gives each participant three colored sticky dots and tells them to place their three dots next the topic or topics they would most like to see covered in the class. Participants may put their dots next to three separate items or put up to three next to one topic. The items that garner the most dots serve as a quick visual aid in determining class needs.

To promote team-building and to move participants to action when they return to work from training, Lloyd Parsons, a staff assistant for employee development at Sears Canada Inc., Toronto, uses the "graffiti wall" exercise:

Break the class into teams of three to five participants. Give each team a length of brown packaging paper or several sheets of flip-chart paper taped together (10 to 16 feet long), different colored markers, and masking tape. Using what they have learned from the course, have each team develop a plan they can put into action back on the job. Tell teams to use pictures, symbols, and words to illustrate that plan on the provided paper. To help trainees visualize the completed "graffiti wall," use the analogy of New York subways.

Give the groups about 30 minutes to finish their graffiti walls, then hang them on classroom or hallway walls. Invite trainees to take a stroll by the walls. Then have each group give a short presentation explaining their work.

Encourage trainees to take their creations back to work and post them in their employee entrance or lunchroom to see what reaction their "walls" get.

93

'Graffiti wall' moves trainees to action back at job

94

Team exercise shows importance of diverse learning styles

Judy Still, a trainer with Healthcare Retirement Corp., Whitehall Manor, MI, uses a game she calls "The Road Less Traveled" to show the importance of diverse backgrounds and thinking styles in reaching team goals.

She hangs a large, laminated map of the United States on the classroom wall, and selects six participants. Each is given a washable marker and a "destination card," and told not to discuss the cards until asked. The cards describe specific travel itineraries and some sights along the way (see example on opposite page). The trips may start from the same or different places, but all must end at the same destination.

Participants take turns going to the map and drawing their travel routes. After all have drawn their routes, Still asks them to tell what specific directions were on their cards and why they selected that route. She also asks them to improvise descriptions of one or two fascinating experiences they had along the way.

She then asks the group to discuss where participants started, where they went, and where they eventually ended up. The message is that people can have varying motives and take differing routes to reach the same goal, and that no particular route is necessarily better than the others. By taking different paths, each person brings knowledge and information to the group that it otherwise might not have. Finally, she asks the group to compare this lesson on diversity to their own management styles.

Sample Destination Card

Your journey begins in _____ and your final destination is _____.

You have always been a great lover of jazz music and just lately you have developed a taste for Cajun food. Your interests will take you to the great jazz clubs of Atlanta, Chicago, and New Orleans. In New Orleans, you get to enjoy the Cajun cooking of Chef Paul.

Identify the significant highlights of your trip for the group.

95

Keeping props under wraps raises trainee interest

A little intrigue can keep participants attentive as you introduce new material. Here are two tricks from Carol Webb, staff development specialist, at the Cooperative Extension Service, Clemson University:

• Arrive early and place any props you plan to use under a tablecloth at the front of the room. Curiosity and interest will be high as you uncover the items one at a time to illustrate your topics.

• Use the "grab bag" approach either to introduce a lesson or to summarize a section. Assemble items representative of the topics you plan to discuss or summarize, then involve the audience by pulling the items from the bag and asking the group how they relate to the lesson. People can chime in with additional ideas or correlations the item brings to mind.

Students in classes given by Craig Hauser and Dale Morehouse, trainers with Walt Disney World, Orlando, FL, get to know one another via "wanted" posters.

Upon arrival, attendees are given a sheet with two empty boxes at the top for front and profile mug shots, and blanks below, labeled "background" (information about family, place of birth, career information), and "known behaviors" (information about hobbies). Other points of interest may also be included to add flavor to the posters, such as "car last seen in," or "known to watch the following TV programs."

The trainer tapes the posters to a wall. Participants are asked to read the posters and mark at the bottom whom they think it describes.

The trainer, meanwhile, asks each participant to pose for two Polaroid photos taken against the backdrop of a height marker similar to actual mug shots. Instead of holding a serial number, participants create tagboard signs displaying their first names and any "known alias," i.e "Nancy a.k.a. The Accountant."

After trainees guess whom the posters belong to, the instructor asks students to tape their photos on their posters.

96

'Wanted' posters help acquaint participants

97

Before- and after- training videos show trainee progress

Film — or videotape, in this case — doesn't lie. So in assertiveness classes, Kelly Miller, manager of training and development at Sallie Mae Loan Servicing Center/PA in Wilkes-Barre, PA, videotapes participants role-playing scenarios before and after training to illustrate their progress.

Prior to the training session, Miller asks participants to submit examples of situations in which they've had difficulty asserting themselves. Some examples include dealing with an irate customer or handling a disgreement with a supervisor.

At the beginning of the session, Miller videotapes participants acting out the scenarios. Before conducting any training, she shows the videotape and asks attendees to write down any assertive, aggressive, and passive behaviors they observe. She tells them to watch for nonverbal behavior tips and tone of voice. Miller then delivers the training.

At the end of the session, she videotapes participants role-playing the scenarios again, this time using their new assertiveness skills. The class once again views the videotape, identifying and discussing the assertive skills.

R ather than creating a standard, copy-intensive training agenda on a standard sheet of paper, Carole Connolly, youth development advisor for UC Cooperative Extension in Fairfield, CA, uses a visual timeline for participants to follow as a training day progresses. For example, to depict a break, she draws a coffee cup, or to show a small group activity she draws a picture of clusters of "smiley face" attendees.

A quick glance at the large flip chart posted in the front of the room lets participants know what's coming next, which is easier than searching in packets of materials for agendas and certainly more visually appealing, Connolly says.

98

Visual timeline maps out training day

99

Flip-chart activity reinforces presentation techniques

John Kane, manager of training and development for Traco Co. in Zelienople, PA, uses this flip-chart activity to help new trainers avoid using mannerisms that can distract a class during a training presentation.

He first places a number of flip-chart easels back to back, with the word MANNERISMS spelled out vertically on each pad. Participants are divided into teams, and each team is positioned in front of a flip-chart easel, blocking its view of what other teams are writing on their charts.

Teams are given 90 seconds to fill in words or phrases of common distracting mannerisms that begin with one letter of the word "mannerisms."

If a team assigns a mannerism to each letter in the allotted time, everyone receives a prize — a small squeeze change purse — as a reminder of one mannerism to be avoided: jingling coins in your pocket.

The lists are then posted in plain view, to be used later as a reference when participants provide feedback to each other during videotaped practice sessions.

An assortment of arts and crafts materials on display during a training segment on food stamp recertification for the state of Maryland sparks the curiosity of Carol Ann Snyder's students. After she finishes her presentation, Snyder asks participants to help themselves to displayed materials to create a visual of something they learned.

She says she has seen great creativity and camaraderie as participants review the material to create a visual. When they complete the projects, participants tell the group about their works of art and display them in the training room.

100

'Art' projects serve as fun, creative review

101

Pie charts help monitor trainee perception levels

Don Valencic, a trainer for Boeing Commercial Airplane, monitors class comprehension with this technique:

He makes a paper puzzle, usually a circle divided into wedges, for each participant. A word or phrase that represents a concept to be presented during class is written on each wedge or piece.

As each student fully understands a concept, he or she tapes the appropriate wedge or puzzle piece on the wall. Valencic only needs to look around the room to see if puzzle pieces are missing to know whether a trainee needs more time or help on a particular concept. At the end of the session, all puzzles should be complete.

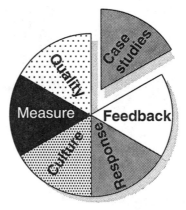

About the Author...

Robert Pike has been developing and implementing training programs for business, industry, government, and the professions since 1969. As president of Creative Training Techniques International Inc., Resources for Organizations Inc., and The Resources Group Inc., he leads over 150 sessions each year on topics such as leadership, attitudes, motivation, communication, decision-making, problem-solving, personal and organizational effectiveness, conflict management, team-building, and managerial productivity.

More than 50,000 trainers have attended Pike's Creative Training Techniques workshops. As a consultant, he has worked with such organizations as American Express, Upjohn, Hallmark Cards Inc., IBM, PSE&G, Bally's Casino Resort, and Shell Oil. A member of the American Society for Training and Development (ASTD) since 1972, he has served on three of the organization's national design groups, and held office as director of special interest groups and as a member of the national board.

An outstanding speaker, Pike has been a presenter at regional and national conferences for ASTD and other organizations. He currently serves as co-chairman of the Professional Emphasis Groups for the National Speakers' Association. He was recently granted the professional designation of Certified Speaking Professional (CSP) by the NSA, an endorsement earned by only 170 of the organization's 3,800 members.

Pike is editor of Lakewood Publications' *Creative Training Techniques* newsletter, author of *The Creative Training Techniques Handbook*, and has contributed articles to *TRAINING Magazine*, *The Personnel Administrator*, and *Self-Development Journal*. He has been listed, since 1980, in *Who's Who in the Midwest* and is listed in *Who's Who in Finance and Industry*.

Want More Copies?

This and most other Lakewood books are available at special quantity discounts when purchased in bulk. For details write Lakewood Books, 50 South Ninth Street, Minneapolis, MN 55402. Call (800) 707-7769 or (612) 333-0471. Or fax (612) 340-4819. Visit our web page at www.lakewoodpub.com.

More on Training

Powerful Audiovisual Techniques: 101 Ideas to Increase the
  Impact and Effectiveness of Your Training $14.95

Dynamic Openers & Energizers: 101 Tips and Tactics for
  Enlivening Your Training Classroom $14.95

Optimizing Training Transfer: 101 Techniques for Improving
  Training Retention and Application $14.95

Managing the Front-End of Training: 101 Ways to Analyze
  Training Needs — And Get Results! $14.95

Motivating Your Trainees: 101 Proven Ways to Get Them to
  Really Want to Learn $14.95

Creative Training Techniques Handbook: Tips, Tactics, and How-
  To's for Delivering Effective Training, 2nd Edition $49.95

Creative Training Techniques Newsletter: Tips, Tactics, and
  How-To's for Delivering Effective Training $ 99/12 issues